The Iron State---Its Natural Position, Power and Wealth.

ADDRESS,

DELIVERED BEFORE THE

NEW JERSEY HISTORICAL SOCIETY,

AT ITS

NINTH ANNUAL MEETING,

Held in Trenton, on Thursday, Jan'y 19th, 1854.

BY HON. JACOB W. MILLER.

NEWARK:
PRINTED FOR THE SOCIETY, AT THE DAILY ADVERTISER OFFICE.
1854.

ADDRESS.

Territorial position is one of the elements of national power, and the geographical features of a country give direction to the labor, and tone to the character, of its inhabitants. Nature, governed by unerring laws, is superior to art, and man, with all his wisdom, must yield to the works of his Creator. Climate and soil, rivers and mountains, oceans and continents, contract or expand the enterprise of man, and control the destiny of nations. History is not always true to nature. It yields to man and his works more credit than they deserve, by representing that the power and prosperity of a country are but the results of association and government, and that national wealth is created by the labor and craft of its inhabitants. Under its flattering teachings we are brought to look upon a land, teeming with the richest productions of nature, and supporting by its bounty millions of free and happy people, as only a work of art, wrought out by political economy, and sustained by the administration of civil authority. In our exaltation we overlook the true sources of prosperity. We forget that national wealth is drawn from material nature; that a genial climate gives health and vigor to population; that the soil which man treads as dust beneath his feet, provides him with food, and clothes him with raiment; that the mountains are store-houses of inexhaustible treasures; that the rivers, the natural channels of trade, give value to the productions of art; that oceans are the great highways of commerce, and that upon the wings of the wind are borne the rich profits which build up and sustain magnificent cities.

Men and governments, war and politics, do not make up all of history. They are but the actors and the shifting scenes in the great drama of national life. Generations of men come and go; social and political institutions, ever changing, rapidly pass away; but the *Land* upon which they lived and moved, remains as fixed as the everlasting hills. And Nature, as bountiful as her Creator, ever exists to supply from her inexhaustible fountains, the wants of men, and to bestow wealth upon the nations.

While it is our duty and our pride to collect and record each old stirring legend and traditionary story, portraying the heroism and virtues of our ancestors, and never to forget the history of those glorious institutions under whose shelter we now enjoy personal and political liberty; let us not be unmindful of the *Land*, that rich inheritance, which God in his providence has given to us for our home and for our country. "It is a good land, a land of brooks of water, of fountains and depths that spring out of valleys and hills, a land of wheat and barley, a land wherein thou shalt eat bread without scarceness, thou shalt not lack any thing in it, a land whose stones are iron, and out of whose hills thou mayst dig brass."

The Territory of New Jersey has a peculiar history of its own. It antedates our political annals. It carries us back into the freedom of Nature, when this broad continent in all its vastness, and wealth, was unappropriated to the use of civilized man, either as private property or public domain. It informs us when and by whom, and under what circumstances, the territory, afterwards called New Jersey, was first selected from the vast domain, and by fixed boundaries, appropriated for separate ownership and government. It teaches us that title to the land preceded the right to govern its inhabitants; that deeds were before constitutions and that private grants gave immunities to property, which survived the prerogative of the royal grantor; and which are still preserved by constitutional law, as the sacred and inviolable rights of freemen. It was by and through these old land title-deeds, that the colonists acquired right to, and fixed the boundaries of, those thirteen several territories, which now constitute the Atlantic States of our Union.

Had the territory of New Jersey been acquired by conquest, and its boundaries settled by border wars; had its mountain passes and river banks been the scenes of bloody conflicts, with the aboriginal owners; then our early history would have been written in traditionary lays, and in legendary song, giving name and distinction to hills and valleys, by their association with the heroic deeds of our ancestors; but a land acquired by fair purchase, with no higher origin than a parchment deed, we must be content to write its annals in humble prose. Yet the history of such a land is not destitute of interest. The peaceful efforts of man, to subdue physical nature to his use, is a contest which approaches the heroic. It is the conflict of labor, single-handed, unaided by capital, and without the facilities of art, making its first onset upon stern nature, and writing its own history upon the soil, in those great landmarks of fallen forests and cultivated fields, and in deep mines of the mountains.

There is a history written by the sword, in blood: there is also a history written by the hand of labor, with the sweat of the brow; the former is preserved by art and poetry; the latter is impressed upon the soil of the valleys, and engraved upon the iron rocks of the mountains, and illustrated by the rich and varied productions of the land. The former may administer to the pride and glory of a people; but the latter teaches us the true principle of political progress, by recording the results of labor and enterprise, as they are developed in the gradual improvements, and in the advancing prosperity of the country.

On the 24th of June, 1664, James, Duke of York, by one of those ordinary instruments, known to the common law for the conveyance of lands, granted the soil, and fixed the boundaries of the country which now forms the State of New Jersey. By the same deed, which conveyed title to the land, was also granted the prerogative of government; and if allegiance could be made the subject of bargain and sale, his royal highness under authority from his brother, King Charles, by a parchment deed of lease and release, gave both territorial identity, and political existence to a State.

The land is still held under the title granted by this deed, and according to its general boundaries; so that the territorial jurisdiction of the State, and the private land-titles of its people are derived from the same source. Thus did the Merry Monarch of England, to gratify a passing whim, or to reward a favorite of the Court, or perhaps to rid himself of an urgent creditor, give geographical position to one of the independent States of our Union, and transferred to the possession of our fathers that goodly land, upon which they afterwards erected those noble civil institutions, which now protect and defend the life, liberty and property of Jerseymen.

This incongruous association of title to the lands, with authority to govern the people, formed by the grants to which I have alluded, is a singular feature in our provincial history. It exerted a controlling influence over the settlements and progress of the colony, and gave a peculiar character to its laws and social institutions. Out of it grew the proprietary government, which for a time divided the colony into two separate political jurisdictions, each with their local government, creating sectional customs and feelings, the influences of which are not yet entirely lost. Yet, in and through this conflicting and confused system of deeds and concessions of proprietary rights and civil authorities, we must trace our territorial, legislative and judicial history, extracting from old deeds, obsolete statutes, and forgotten decisions, many a principle of law and government, which now gives security to our property, and protection to our liberty. This task has already been ably performed by two worthy members of this Society, and the results of their labors given to the public in those two most valuable volumes, "East Jersey under the Proprietary Governments," and "The Provincial Courts of New Jersey."

The political portion of the grant was destined to a different fate. The contract conveying the property in the soil, survived the grant of civil government over the people. While neither war nor revolution disturbed the *title* to lands derived from the king, his authority to transfer the allegiance of his subjects, although acquiesced in for a time, was never acknowledged by freemen, and was finally scattered to the winds, as a baseless assumption of power, in the storm of the revolution.

The royal grantor lost his throne, and died in exile. And the American people, on the very soil which he granted, wrested by war from the hands of his most powerful successor, all right of government over the territory; yet the parchment deed of James, Duke of York, survived both events, and still exists a respected muniment of title to all lands within the boundaries of New Jersey. Thus, we see written in our territorial history, and recorded with our title-deeds, that great conservative principle, now the organic law of the Union,—the sacred right of private property, and the inviolability of legal contracts.

It should be noted in our annals that the territory of New Jersey was acquired by purchase and not by aggression; that it was neither wrested from the Crown of England, nor stolen from the Indians; but that every acre of the wide domain is held by legal tenure, unstained by blood, and unpolluted with fraud. It is to this land, thus acquired, and thus holden, that I would call your attention.

The Territory of New Jersey—its natural position, power and wealth; these, together with some old chronicles, rescued from the impending rubbish, under which time buries memory, shall be the subject of my address, and will, I trust, be an appropriate offering to the historic Muse who presides over our Institution.

The POSITION OF NEW JERSEY is one of great natural beauty, and of immense power and influence. With the ocean in front, and flanked on either side by two noble rivers, her territory, well defined and defended, stretches from the sea-board to the blue ridge. To the north, are mountains clothed in forest and rugged in rock, but these rocks give covering to endless accumulations, preferable to the mines of Potosi and Golconga; for zinc, copper and iron are masters over gold and diamond, and create the highest of all productions, the strong arm, the bold heart, the energetic enterprise of freedom. Between the mountains, are vallies waving with the benediction of

Ceres, and orchards bending under the luscious gifts of Pomona, while industry seats herself upon every rill, and the air is vocal with the sounds of the hammer, the wheel, and the axe. To the south, the land assumes a less rigid aspect; meadows stretch far and wide, and the pastoral scenes of Acadia are renewed. On mountain, valley and meadow, the house of prayer forms a leading feature in the landscape, and from every town and village, tall spires point to the upward path of pure morality and fervent piety, while court-houses proclaim the empire of law, and numberless poles announce the universal reign of liberty.

Our position thus secures to us the advantages of foreign commerce and the facilities of internal trade. It is true that our two powerful neighbors have almost monopolized the first, but they cannot deprive us of our locality. And New York and Philadelphia, although outside of our boundaries, afford to us the two best markets in the country. The barrel may be tapped at both ends, but *the barrel* is still *our own*, to be filled or emptied at our pleasure, according to the demand and the supply.

Our geographical situation is also one of national consequence. I speak not of the political influence, which extent of territory and population may give in the administration of the general government, but of that power which accrues to a state from its connection with the great events of history, from its position in the struggle for liberty, in the march of empire, and in the development of national wealth. It was not accident, that made New Jersey the "Flanders" of America; it was not choice, that made her soil the battle-field of the revolution; it was the central position of her territory which exposed it to the shock of war, and for a time converted it into a broad highway for the tramp of armies. The position which she held in war, she is entitled to enjoy in peace. She still controls the road from New York to Philadelphia, over which in 1776, the car of war rolled from the Hudson to the Delaware, but now the great thoroughfare of travel and of commerce, over which two great cities daily transport their merchandise, and nightly send a portion of their weary population for repose among the green hills of Jersey. And along that line of battle-fields, where armies fought for the empire of America, now peacefully pass the rich and varied products of thirteen free States, whose independence was *there* achieved. Our territory lies between the ocean and the coal-beds of Pennsylvania, and also commands the shortest line of travel from the western lakes. And at this moment, vast accumulations of products from mines, land and forests, are pressing upon our western boundary, for a right of way over our soil to the markets on the seaboard. This right of way is part of our great freehold estate, is incident to the public domain, and belongs to the people of New Jersey in their sovereign capacity.

The roads of a country constitute part of its wealth. They increase productions, stimulate labor and enterprise, and give facility to trade and business. Modern invention has given to local roads. a more extended use, and clothed them with a national character. Commerce, formerly confined to ships, and conveying its wares upon rivers and oceans, now with the steam-engine claims empire over the land, and upon the railroad track sends mighty cargoes of merchandize, and armies of men from city to city and from State to State.

The territory of New Jersey forms a controlling section in those great inland ways of commerce. And this control over the right of way is ours for the great purposes of trade, and for the general welfare; it is ours for improvement and progress, for the development of the natural wealth of the land, and for the encouragement of the labor and enterprise of its people. But while it is our privilege to possess this right, and our duty to improve it for the benefit of our own State, let us not forget that it is also an instrument committed to our hands, for the advancement of the power and prosperity of the whole Union. Let us remember that our railroads are but links in that great chain of internal improvements, which, stretching along the whole extent of the Atlantic coast, and passing far out toward the west, is soon destined to reach the shores of the Pacific, binding together states and cities, towns and people, by the strong ties of social and commercial relations, and uniting all in a closer union of national feeling, interest and sentiment.

New Jersey has done much for internal improvement, and our roads and canals, constructed under the authority of our Legislature, by the enterprise and capital of our citizens, compare favorably with those of our sister States. Yet we have not improved the advantages of our position to their full extent; we have looked more to present profit than to future advancement, and our system partakes more of the character of private business, than of public enterprise.

With the advancing prosperity of the country, and the increasing demand for new facilities of travel and transportation, our position will enable us to build and maintain the most useful and the most profitabe roads upon the continent. These great thoroughfares will not only give to New Jersey a commanding influence over the internal commerce of the whole country, but will also enable her to improve to their full extent, her own rich, but neglected fields, of mineral and agricultural wealth, by giving to every farm-house, and forge, and workshop, the inestimable advantage of rapid and cheap transportation, by making every mountain pass echo with the roar of the passing trains, and by bringing each homestead from Cape May to Sussex within the sound of the locomotive's whistle. For these advantages, when we enjoy them, we shall be indebted to our territorial position, to be secured and improved, or wasted and lost, according to the policy we may adopt.

But New Jersey has other treasures besides that of the right of way over its surface. With many varieties of soil and exposure, warmed by a sun whose temperate rays render the climate neither too hot nor too cold, and refreshed by clear cold mountain breezes, tempered by the milder air of the ocean, the land produces all kinds of useful grain, brings forth the earliest products of the spring, and ripens the most luscious fruits of autumn, while the variegated scenery, and accessible elevations, invite the stranger to take up his abode upon the hill-side or in the valley, and bind the heart of every Jerseyman to the home of his fathers. Nature has dealt bountifully with us; there are also treasures beneath the soil, and where the rocks forbid the plough, and where the sterility of the land rejects vegetation, there, beneath the barren and rugged surface, in deep broad veins are deposited vast treasures of rich and valuable minerals.

A range of rough and broken hills, extending through the counties of Passaic, Sussex and Morris, and passing off into Warren and Hunterdon, mark the rich mineral regions of New Jersey. These hills are made accessible on all sides, through gorges and valleys, formed by numerous mountain streams, the head waters of the Passaic, the Raritan, and the Musconetcong, which afford by their perennial supply and rapid descent, a cheap and never-failing water power.

But a few years ago, the traveler in passing up and down these water courses, would have seen little to attract his attention, except here and there, a forsaken excavation in the mountain side, disturbing its rugged soil by fragments of up-torn rock, and old ruins of massive stone walls, surrounded by heaps of cinders, marking the spot, on the banks of some stream, where forge and furnace in times long gone by, converted the ores of those mountains into iron and steel. These old mines and forges have traditions of their own. Their discovery, the character of those who first opened and worked them, their early progress and success, their subsequent falling off and failure, the manner in which their advantages were appreciated in war and neglected in peace, would form a most useful chapter of political economy, teaching by time and experience that which theories cannot anticipate. I have neither the time nor the ability to perform this task, but I must content myself with gathering a few of the broken fragments, which lie scattered upon the surface of this unworked mine of historic lore.

About the beginning of the eighteenth century, when the manufacture of iron was yet in its infancy, before the mines of Wales had felt the effect of English capital, when Birmingham was yet a town of little note, and Liverpool just rising into commercial importance, a vessel from a foreign port discharged a small cargo upon one of the London docks. The quality of the importation attracted the attention of mechanics and ship-builders, and its superiority over everything of the kind then manufactured in England, was noticed by all. This strange but appreciated cargo, was *Jersey Iron*, made from ore dug from the mountains of Sussex. The mine, which supplied this first importation of American bar iron into England, still exists, and although long neglected, has lately been re-opened, and under the direction of its enterprising owners, again yields its rich ores for the use of the country. This mine is a type of many others of equal importance, and its history will illustrate the progress of the iron business of New Jersey, through years of changing success and adversity, up to its present improved condition.

Fifty years after the grant of the territory of New Jersey, by the Duke of York to Carteret and Berkley, and when the House of Stuart had ceased to reign over England and her American Colonies, a man who had been a favorite at the court of the dethroned monarch, but uncorrupted by its vices, became a freeholder in New Jersey. He was a statesman, and a philanthropist. His name is perpetuated by one of the largest and richest States in the Union, and his principles of moral and social government were deeply impressed upon one half of our State. And although his mission to America was the great work of establishing upon the New Continent an empire of peace, liberty and law, he was not unmindful of the natural wealth of the country, and the lands of New Jersey did not fail to attract his attention. On the 10th of March, 1714, by a warrant from the Council of Proprietors, he acquired title to a large tract of land, situated among the mountains, then of Hunterdon, now of Sussex county, and WILLIAM PENN became the owner of one of the richest mines of iron ore in New Jersey. This mine, since called Andover, was opened and worked to a considerable extent, as early as 1760. Forges and furnaces, the ruins of which are still visible, were erected for smelting ore, and making it into bar iron. Tradition reveals to us, that the products of these works were carried upon pack-horses and carts down the valley of the Mosconetcong, to a place on the Delaware called Durham, and from thence transported to Philadelphia in boats, which were remarkable for their beauty and model, and are known as Durham boats to this day.

At the breaking out of the Revolution, the Andover Iron Works had acquired sufficient importance to command public attention; before that period most of their iron had been exported to the mother country, and there used for government purposes. But now steel and cannon balls were required for the use of the confederated Colonies, and the iron ores of New Jersey were to be put under requisition for the defence of the liberties of the people. "And the Andover Iron Works were ordered to be put in blast, for the purpose of procuring iron to be made into steel, it being represented that Iron made at the said works, is the most proper of any in America for that purpose." But unfortunately for the public service, Andover was under the control of the enemy. Its owners were enjoying the protection of the British army in Philadelphia, and all its iron had been converted into hostile steel. This emergency produced the following resolution:

"*In Congress of the Confederation of the*
"*United States of America.*
"Thursday, January 15th, 1778.

"The Board of War brought in a report; whereupon, resolved, That the "Board of War be authorized to direct Colonel Flower to make a con-"tract with Mr. Whitehead Humphreys, on the terms of the former "agreement, or such other as Colonel Flower shall deem equitable, for "making of steel, for the supply of the Continental Artificers, and works "with that necessary article; and as the iron made at the Andover "Works only, will with certainty answer the purpose of making steel, "that Colonel Flower be directed to apply to the Government of New "Jersey to put a proper person in possession of these works, (the same "belonging to persons who adhere to the enemies of these States) upon "such terms as the Government of the State of New Jersey shall think "proper; and that Colonel Flower contract with the said person for such "quantity of iron, as he shall think the service requires.

"*Resolved*, That a letter be written by the Board of War, to the Gov-"ernor and Council of the State of New Jersey, setting forth the pecu-"liarity of the demand for these works, being the only proper means of "procuring iron for steel, an article without which the service must irre-"parably suffer; and that the said Governor and Council be desired to "take such means as they shall think most proper, for putting the said "works in blast, and obtaining a supply of iron without delay."

New Jersey promptly answered this call, on the 18th of March 1778, by the following legislative resolution:

"The Council, having taken into consideration the resolution of Congress "of the 15th of January last, and the letter from the Board of War ac-"companying the said resolution, recommending it to the Government "of this State to cause the Andover Iron Works in the county of Sussex, "to be put into blast, for the purpose of procuring iron to be made into "steel; it being represented that the iron made at the said works is the "most proper of any in America for that purpose: And having also taken "into consideration, the application of Colonel Benjamin Flower, Com-"missary General of Milliary Stores, agreeably to the said resolve, who, "at the same time recommended Colonel John Patton, as a proper person "to carry on the said works: And considering, that it is not yet ascer-"tained that the estate in the said Andover Iron Works is confiscable to "the use of the public, or whether the owners thereof have committed "any act of forfeiture; and at the same time being desirous that the pub-"lic service may be promoted, by the use of the said works;

"*Resolved*, That it be recommended to Colonel Patton to agree with "the present owners of the said works to take the same, to wit: the fur- "nace and forges on lease, hereby assuring him, that in case the said es- "state shall be legally adjudged to be forfeited, or otherwise become un- "der the particular direction of this Government, such agreement shall "be confirmed to the said Colonel Patton, or to such person or persons "as the Legislature shall approve, for any period not exceeding three "years from the date hereof: But if the said owners shall refuse to let "the said works for the use of the public, the Legislature will then take "the necessary steps for putting them in the possession of a proper "person in order to have them carried on for the purpose above men- "tioned.

"*Ordered*, That Mr. Hoops wait on the House of Assembly, with the "foregoing resolution, and desire their concurrence therein."

"Which message being read and considered, Resolved, That the House do concur in the resolution contained in the said message."

Under these authorities the old Andover Works change owners. Passing from the control of their traitorous proprietors, they are now in the hands of true men for the use of their country, mine, and furnace, and forge seem to catch the patriotic spirit of their new occupiers, the fires glow with an intenser heat, and the anvils ring louder and clearer, as if conscious that they are forging arms with which brave men shall defend their homes and their country.

It was not only Andover that responded to the call of the Government for aid, but all along that mineral region, from Sterling forge in Bergen, to Union furnace in Hunterdon, was one stirring scene of action, effort and labor; miners and forgemen, wood-choppers and colliers, urged on by citizen soldiers and patriotic officers, were all engaged in procuring iron and steel for the use of the Continental army, while through the valleys and the gorges came the echo of the sound of the hammers, as, swung by stalwart arms, they rang upon the anvils, and kept time to the song of the forge.

"Clang, clang! the massive anvils ring;
Clang, clang! a hundred hammers swing; . . .
. . . Say, brothers of the dusky brow,
What are your strong arms forging now? . . .
The Sword!—a name of dread— . . . .
. . . Yet still whene'er the battle-word
Is Liberty, when men do stand
For Justice and their Native Land,
Then Heaven bless the Sword!"

War had made terrible ravages in New Jersey; her brave men had been slain in battle, her towns had been sacked, and her churches and farm-houses given to the flames; her State treasury was bankrupt, and her people impoverished, yet, thank God! her means for the defence of liberty and country were not yet exhausted; her mineral wealth was beyond the reach of invading armies, and her iron mines entrenched in rocks, defied the power of England. And now, at the call of liberty, out of the deep caverns of the mountains, as from a mighty arsenal, pours forth the true metal of war, iron and steel, and New Jersey, in the hour of her country's utmost need, furnishes both the soldier and his sword.

But to continue our history. The Andover Works were held by the Government until the close of the war, and the mines of New Jersey for five years furnished iron and steel for the Continental army. Then came peace and independence, and the country, rejoicing in its fresh liberty, soon recovered from the devastations of war. The land, relieved of hostile armies, again yielded its rich harvests of grain and fruits to reward the labor of the husbandman; and commerce, young and lusty, plumed her white wings over the free ocean, and commenced that onward flight, which has since borne our ships to every sea. But there was one interest in the country, which did not partake of this reviving prosperity. That which had by its national importance commanded the attention of Congress and State Legislature, is now neglected and forgotten; and the manufacture of iron from native ores, which was found to be so invaluable for the defence of the country in war, is not thought of sufficient consequence to demand the encouragement and protection of the Government in time of peace. It is true, that for a few years after the peace of 1783, that great interest, owing more to the state of our foreign commerce, than to any efficient domestic aid, continued to advance, and the iron mines of New Jersey were worked to adtage by their owners. But this short-lived prosperity was soon to be followed by long years of adversity. Left to contend with their old enemy in a new field of warfare, the iron mines of New Jersey, which had repelled the armies of England, fell before her invincible capital. And now the scene again changes at old Andover; forsaken by the Government, its owners driven off by bankruptcy, its mines deserted, its furnaces having given their expiring blast, and its forge-hammers resting upon the anvils, nothing but heaps of ruins marked the place, where labor and enterprise had once supplied the wants of a nation.

But the war of rival interest did not stop here; the enemy was not satisfiied with the temporary destruction of our iron works, and the bankruptcy of their owners. The inexhaustible mines were still there, to supply materials for a renewal of the contest at some more propitious moment. And advancing upon the ruins of our prostrate manufactories up to the sources of our national wealth, England went on until she was enabled to imprison our ores by the iron bars of Wales laid across and over the very doors of our mines.

Advantages in commerce as well as those of war, when pushed to extremities, produce reaction. This triumph of English iron over American, was too destructive to our interest, and too humiliating to our national pride to be long submitted to.

The prosperity of the country outstripped the cautious policy of the Government, and individual enterprise and private capital, stimulated by the wants of trade, came to the rescue, and manfully contended with English manufactures against their monopoly of free trade in iron. Now, all again is life and activity in those neglected mineral regions. American labor and enterprise, with strong arm and bold hand, with railroad and canal are there, contending with the might of British capital.

They have stormed those mountain heights, and unbarred the doors of the imprisoned mines; and agian the emancipated ores come forth in triumph to the music of an hundred forges, and American iron once more successfully competes with the English manufacture.

Those sterile mineral fields are again occupied, and feel the effects of labor and capital. But together with the miner and the bloomer come a host of strange operators. They are workers neither in ore nor in iron; yet they are laboring in digging deep excavations in those hills, and in making railroad contractors, with their army of Irish laborers, have entered the broad highways leding into the very doors of the mines. Engineers and mineral field, and constitute part of the efficient force employed in modern mining. The progress of the arts has developed new elements of power, for the raising and smelting of ores, andfor the manufacture and transportation of iron.

The steam-engine becomes a mining instrument, and relieving the tedious labor of man and horse, lifts the ponderous mineral from its deep damp bed, and then, sends it with mighty speed upon iron ways into distant localities, to meet the coal and assimilating ores of Pennsylvania, to be converted into iron beneath the hot blast of the furnace.

That revolutionary mine, for fifty years a neglected waste, has been transformed by the magic power of modern art, into a deposit of productive wealth more valuable than gold, and has sent, during the last five years, upon railroad and canal, constructed along the line of that old cart-way, 150,000 tons of its rich ores to the banks of the Delaware. And by a strange coincidence, upon the very site, where in 1778, these ores were converted by the Government into weapons of war, the present proprietors of the mine, have erected their works, from which they daily roll out tons of iron, for the implements of husbandry, the tools of mechanical arts, and tho great pathways of commerce.

It is not within the scope of this address, to enter into an analysis of the iron ores of New Jersey. My object is merely to call public attention to their admitted superiority, so that their importance as a source of wealth and prosperity to the State may be more fully appreciated. A few statistics will accomplish this. Within the mineral region to which I have alluded, there were raised during the last year about 175,000 tons of ore. New veins of rich and extensive deposits have lately been discovered, and are now in process of being opened; these, with the improved facilities of mining, stimulated by the advancing demand for iron, will, it is estimated, increase the production of our mines during the coming year to 250,000 tons.

I will give you a still more practical illustration of the increasing value of our mineral productions. In the year 1851, one of the largest iron manufacturing establishments in the county of Morris was compelled, by the ruinous state of the iron trade in this country, to undergo the mortal process of a sheriff's sale. In the hands of its new owners, and under a more auspicious state of the market, its fires were re-kindled in 1852, and during the last year "Boonton Iron Works" used 11,600 tons of Jersey magnetic ore, consumed 23,000 tons of Anthracite coal, 3,000 tons of limestone, 6,600 tons of pig iron, employed in its operations 600 men, paid out for wages $22,000 per month, and manufactured 6,500 tons of nails and railroad spikes. Other establishments in the State consume a still larger quantity of ore, while the demand from abroad is daily increasing.

Our mineral productions are also about to be enlarged, by the opening and working of extensive veins of the Franklinite. This ore, by reason of its peculiar combinations, has hitherto been of little use in the manufacture of iron; but nature's concretions, although not readily comprehended by man, are always intended for his benefit; he has only to discover the key which will unlock the mystery. The discovery has been made, and this salamander of the charcoal furnace, now yields to the heat of the Anthracite, and becomes both a flux and a vapor, producing the best of iron and the most durable of paints.

In the year 1852, about one hundred years from the time when that first cargo of Colonial bar iron made its appearance in England, there was placed at the *door* of the Crystal Palace in London, because it was too large for entrance, a mineral rock, which by its size and rare quality, commanded attention even at the World's Fair. This was a Jersey production, a pebble specimen of our mountain of zinc. And the New Jersey Zinc Company had the honor of obtaining the prize medal, over the competing companies of France and Belgium. The Committee which awarded this prize, composed of the most distinguished chemists, pronounced the introduction of the oxide of zinc as a white paint in place of salt lead, as one of the remarkable events in the recent history of chemical art.

This new use to which zinc ore is now applied, will soon make it one of the most important of our minerals. The New Jersey Zinc Company, the first of the kind in this country, commenced its operations about three years ago. In 1852 they manufactured 1,200 tons of paint. In 1853 they raised 6,982 tons of ore, producing 2,200 tons of paint. Their improved works are now making regularly from 75 to 80 tons of paint per week, and during the present year, they expect to mine 12,000 tons of ore.

If experiments now being made prove successful, our zinc ores will also assume national importance, by affording the only chemical substance which will protect our naval and commercial ships against the ravages of those destructive agents of the sea, the marine worm and barnacle formations.

This Jersey manufacture has also acquired a celebrity seldom attained by an American production. It not only embellishes the rooms of our Democratic houses, but has found its way into royal palaces, and it is said that one of the apartments in Windsor Castle may be distinguished from the others, by the glossy whiteness, which is peculiar to Jersey zinc paint.

New Jersey with her magnetic and Franklinite iron mines, and her zinc deposits of inexhaustible supply, possesses the richest field of mineral wealth in America; these, together with the advantages of her location, secure to the State a source of public wealth, to be limited only by the uses to which iron and zinc may be applied by the art of man.

The extent to which this source of wealth may be improved for the advancement of national prosperity and power, is seen in the result produced by English labor and capital, upon a mineral field not much larger nor richer than the one which it is our privilege to possess, and our duty to improve.

"At the close of the reign of Charles II," says Macaulay, "a great part of the iron which was used in the country, was imported from abroad, and the whole quantity cast here annually, seems not to have exceeded *ten thousand tons*." In 1740 it had only increased to 17,000 tons, and 59 furnaces constituted the whole of the iron works of the kingdom. From this period the iron manufacturers of England commenced their onward march. The revolution of 1688 had relieved the working classes from the crushing burdens of unlawful exactions, and liberated capital from the grasp of royal monopolies. Industrial pursuits became more honorable and more profitable, and the mechanical arts, stimulated by the pressing demands of the country, come to the rescue of the neglected manufactures. Then came American Independence, which, depriving England of her Colonial resources, compelled her to look to her own mineral fields for the means of national prosperity. Nobly has she improved those fields, under the protection and encouragement of the Government, by labor, by invention, and by capital, until their rich and multiplied productions command the iron market of the world. Occupying this advanced position, the iron manufacturers of England, were prepared to take advantage of the progress of events, and to turn to their own account those mighty discoveries in the arts which distinguish the present age. The knowledge of smelting ores with stone coal, the discovery of the hot blast, and the steam-engine, the modern system of railways, were inventions and improvements for the benefit of England, and from which her manufacturers received the first and the richest profits.

In 1740, the fifty-nine neglected forges of Great Britain produced only 17,000 tons of iron; in 1852 her thousand protected furnaces and mills rolled out about 3,000,000 of tons. Prior to 1776, England imported bar iron from her Colony of New Jersey, to supply her home market. In this seventy-eighth year of our independence, she exports to the United States 500,000 tons of manufactured iron.

The mines of Great Britain, in their present high state of improvement, constitute the most productive source of national wealth in the world. The iron of England, more valuable than the precious metals, now commands the gold of California, and by its essential uses, has become a medium of exchange, and an agent in the commercial transactions of the world; dealing in government bonds, and in public stocks, it makes States its debtors, and lays empires under contribution; more progressive than the Government, it has come to the aid of the liberal movements of the age, and is an essential element of progress and reform, and that triple alliance of Iron, Labour and Liberty, is rapidly changing the physical and social condition of the world.

How strangely, yet how certainly, does NATURAL WEALTH, God's gift to man, connect itself with the affairs of the world; being itself an element of national power, it impresses its own influence upon the very springs and sources of social and political life, enters alike into the business of men and the policy of Governments; makes itself felt in peace and in war, rouses the stagnant energies of old nations, quickens the dormant life of new countries, and gives direction to the commerce of the world, by furnishing materials for the construction of railroads, stretching from ocean to ocean, and lines of steamers, passing from continent to continent.

This element of POWER and WEALTH so triumphantly developed by England, is also possessed by New Jersey. Heretofore we have not had the ability to improve it. And our iron mines have neither advanced the fortunes of their owners, nor the prosperity of the State. But the time has come when Yankee enterprise can compete upon more equal terms with English capital, and American labor now enters the mineral field with higher prospects of success. Commencing our mining operations just at the time when the progress of society requires the largest development of the natural resources of the world; and when our own country especially, by its wide extending territory and increasing prosperity, is opening the largest and most active market for all the mineral productions, and when by the improved state of the arts, we are enabled to avail ourselves to the greatest advantage of all the new discoveries of the age; the manufactures of New Jersey must advance to the highest point of prosperity, and our State become what Wales is to Great Britain, THE IRON DISTRICT OF THE UNION.

I have shown that our iron mines are associated with our history from colonial times to the present day; how they entered into the Revolutionary struggle, and rendered essential aid in the achievement of national independence. They have given historical character to the State in the past; they still exist to secure her future prosperity. In peace and in war, iron has been associated with the name of New Jersey, and is still an element of her power and wealth. Mountains and rivers, trees of the forests, and the rocks of the quarries, have by their localities and uses, given familiar names to States, and distinguished the escutcheons of nations.

The sons of New York rejoice in the title of EMPIRE STATE, Pennsylvania in the KEYSTONE, South Carolina the PALMETTO, and Ohio the BUCKEYE; and almost every member of our Union has its own peculiar war-cry, from the granite rocks of Massachusetts, to the grizzly bear of California; yet Jersey-men have no kindly name of affection for their native land—no appellation beyond the stiff, formal, officially stereotyped "New Jersey." It would be worthy of the Historical Society—nay, it would seem a province of its appropriate duty—to stamp upon the country of whose annals it is the guardian, some characteristic and descriptive, some familiar and yet dignified, some short and pithy word, by which her sons may hail their mother. A popular word of endearment in the field, the forge, the meadow, the factory; a word of active zeal in the halls of legislation, a word which shall mingle with the crash of artillery and the clash of bayonets, on some stern battle-field where our descendants may struggle for Liberty. Might not such a word be found,—ought we not to salute and present her to her Associate Sisters as THE IRON STATE,—a name indicative of valor, strength, perseverance, industry, and union?

Having discoursed so largely upon mines and minerals, my time will permit but a passing notice of the AGRICULTURAL RESOURCES of the State. Varieties of soil, genial climate, cheap manures, and ready markets, are the chief elements of prosperous husbandry. The people of New Jersey enjoy all these advantages, in the formation and position of the country which they occupy. The loam formations of our Southern counties, the gravel of the Middle, and the clay of the Northern, afford the surest and most available farm land for the production of all varieties of grains and grasses. And in those less favored localities, where the soil has not sufficient strength to produce wheat and corn, the sand-fields are converted into gardens of vegetables, and the hill-sides are covered with orchards laden with golden fruits.

Our proximity to the two best markets of the country, gives great value to our horticulture, by securing a ready and rapid sale of those productions, which by their delicacy and perishable nature, require immediate consumption. Although we may not be able to compete with some of our sister States in corn and wheat, yet our position secures to us the control of the fruit and vegetable markets of New York and Philadelphia.

Besides the inexhaustible lime quarries of Hunterdon and Warren, we have a source of agricultural improvement, somewhat peculiar to the State. Almost the entire sea-board of New Jersey has undergone a geological revolution. The bed of the Atlantic has been raised, and become an arid sand-bank, old enough to shoot up into pine barrens, but not sufficiently old to accumulate an available coat of vegetable mould. Here kind Providence has gathered up for use, beds of rich marl, which, judiciously spread, converts the blank and bleak sterility into a smiling expanse of verdure.

For many years, farming was a secondary object, pursued by men of small means and with limited opportunities; but at present, thanks be to Agricultural Societies, more general education, pecuniary ease, and some notion of machinery and chemistry, as useful co-operators, our farmers are rising in the scale of liberal and successful industry. The hand of improvement, guided by intellectual and pecuniary power, becomes everywhere apparent; our only want is increased means of intercourse and conveyance, to make our State the most productive agricultural district in the Union.

Nations, like individuals, have their golden opportunities; occasions when advantages must be improved, or lost forever. The people of New Jersey now occupy that position. All about us is activity and development; the progressive spirit of the age seems to have touched the very springs of industry, giving new impulses to private and public enterprises, and advancing both individuals and communities to a higher grade of prosperity.

Labor, always an efficient instrument of national wealth, has advanced to be an intelligent agent; adding, to the physical, moral force, it has become the great motive power of the age, in advancing the civilization of the world. The working classes, no longer the mere instruments of capital, and the servants of trade, have become the principals in business, and masters in all industrial pursuits. Private enterprise has come to the aid of the sinking fortunes of States, and free labor boldly enters those fields of national wealth where timid governments feared to tread.

In this mighty industrial conflict, the people of New Jersey, to whom toil is more of a passion than a burden, should hold a front position. Though occupying but a small territory, they have the largest means of improvement, productive lands, rare minerals, accessible markets, commanding highways, noble institutions of learning; these, with free labor and free schools, are the only instruments required by a free and virtuous people, to make their State equal, in importance, the largest of the Union.

The territory of New Jersey, glorious in history, rich in mineral wealth, beautiful in scenery, and healthful in climate, is now occupied by four hundred thousand American citizens, who are the owners of its soil, and the masters of its government; free men, whose labor and enterprise accumulate public wealth, and whose votes direct its administration. Free, yet frugal; independent, yet submissive to law; and using liberty without abusing it, unencumbered by public debt, the people of New Jersey enjoy that happy medium state, which secures them against the corruptions of wealth, and the temptations of poverty, they should be neither the slaves of mammon nor the tools of politicians. To what a lofty social and political pre-eminence may not such a people advance such a State! They may not only make it teem with the richest productions of the field and the mine, of the loom and hammer, but also, out of their abundance adorn the land with the noblest works of art, embellish it with all useful institutions of learning, and sanctify it with beautiful temples of religion.

The increasing prosperity of the State indicates, that we are upon the onward march, yet we still lag far behind some of our sister States. Let us, then, quicken our energy, rally our forces and press forward, and never rest till we place New Jersey in the relative position which she occupied in 1776, in the front rank of the Atlantic States of the Union,—the flag of The Iron State waving as high as the highest.

www.ingramcontent.com/pod-product-compliance
Lightning Source LLC
LaVergne TN
LVHW020644110826
845149LV00004B/1352

* 9 7 8 1 4 1 8 1 9 0 3 6 1 *